Artists as Inventors. Inventors as Artists.

SCOTT GUTHERY

9 8 7 6 5 4 3 2

Contents

List of Figures

List of Tables

The creative scientist needs an artistic imagination.

MAX PLANK

Chapter 1

Artists & Inventors

Neither art or science are front-of-mind as we go about our daily lives. It is rather curious, then, that when you put them together you get something that is very much a part of daily life—the iPhone. Today the combination of art and science is called technology. Two centuries ago when the Athenæum was founded—and before the word 'technology' had been given its current meaning—the synergistic combination of art and science was called the useful arts and the men and women who practiced the useful arts were frequently referred to artists.

Consider for a moment Boston's Zakim bridge. A bridge must, of course, support its own weight and the cars that cross it. It has to do so on a sunny Sunday afternoon as well as during a rush hour Nor'easter. And it has to do this as economically as possible. But we also insist that a bridge be pleasing to the eye. That it just looks as if it's going to get us to the other side safely without a second thought. One of the architects of the Zakim Bridge, Christian Menn, was of the view that economy, serviceability and safety of a successful bridge would revolve around its aesthetics. Get the art right and the science will come along.

Or take another example. There was a junk drawer full of abandoned portable MP3 players before the iPod came along. Steve Jobs found the right combination of art and science in the iPod and swept all the others away. The iPod didn't do anything that its predecessors couldn't do. It just did those things in an aesthetically pleasing way; pleasing to the eye, the ear, and the touch.

Today's tour is about this sort of intersection of art and science, the useful arts. In particular, it is about artists who invented things and about scientists who went about their work artistically.

1.1 Samuel Finley Breese Morse

Samuel Finley Breese Morse (1791–1872) was a portrait and history painter. He was born in Charlestown on April 27, 1791. His father was the famous—or infamous depending on you look at these things—Boston historian, Jedidiah Morse. Morse fils prepared for college at Phillips Academy and in 1810 graduated from Yale where he studied mathematics with Jeremiah Day and the physical sciences with Benjamin Silliman. While at Yale he earned his keep by painting portraits for his fellow students ... $1 on canvas or $5 for an ivory miniature.

Morse was granted his first patent when he was 26 years old. The patent, X2843,[1] was issued on November 3, 1817, was joint with his brother Sidney and Samuel Findley for a fire engine pump. Five years later, in 1822, Morse invented a marble-cutting machine that could carve three dimensional sculpture in marble or stone. He discovered that it is not patentable because it infringed on a patent X3131 issued to Thomas Blanchard on September 6, 1819. These two inventions, along with a similar invention of Hiram Powers discussed below, could be considered the forerunners of today's 3D-printers.

In September of 1819, Morse was commissioned to paint a full-length portrait of President James Monroe[2] by city of Charleston. Morse, then living in Charleston, journeyed to Washington to paint a life bust of Monroe that he carried back to Charleston to turn into the full-length portrait. There are a number of copies of the life bust, the one on display in the Athenæum's Long Room and another hanging in the Blue Room of the White House. The copy in the White House is thought to be have been painted by Morse for Monroe's daughter. Of the copy in the Athenæum's collection, David Dearinger, the Athenæum's Director of Exhibitions and Susan Morse Hilles Senior Curator of Paintings & Sculpture says:

> Morse experts do not believe that it is from Morse's hand, but it certainly is of nineteenth-century origin. The jury is still out. Interestingly, the painting was on extended loan to the US State Department for many years starting in the 1960s and hung during that time in the Diplomatic Reception Rooms. It was finally returned to the Athenæum, at our request, about ten years ago.

---

[1]Patents issued from July 31, 1790, when the patent office was opened until July 1836 when a much more stringent patent law was passed are called X patents. The numbering was not restarted as some believe because of the disastrous patent office fire of December 15, 1836.

[2]Monroe is the individual clutching the flag right behind General Washington in the famous painting by Emanuel Leutz, *Washington Crossing the Delaware.*

The use of electricity to send messages was by no means a new idea when Morse turned his attention the topic in the 1830's. In 1809, a German physician, Samuel Sommerring, for example, used electricity to create bubbles associated with each letter of the alphabet at the receiving end. Sommering's device had a number of drawbacks not the least of which was that you needed to string 24 wires from the sending end to the receiving end and somebody had to be awake and paying attention when the bubbles started bubbling. There is, granted, something aesthetically pleasing about bubbling water but in this instance good science and good art did not make good technology.

Initially, Morse sent numbers from one end to the other using a very simple code for each digit; one mark was a one, two marks were a two, and so forth. The received multi-digit number was then associated with a word using a very large decoding dictionary; 82475 might be "afternoon," for example. It was Morse's partner, Alfred Vail, who came up with the idea of what we now know as Morse code.[3] This let one send letters as just easily as numbers and did away with the dictionary.

Morse painted a portrait of Vail's parents. Finished in December of 1837, it was to be the last canvas portrait that he would paint. The science community tends to think that Morse set down his brush to focus all of his attention on his experiments with electricity. The art community cites the fact that he was passed over twice for a commission that he dearly wanted, a panel for the Rotunda in the Capital. It is probably the case that push and pull were both at work.

Morse was granted US Patent 1,647 on June 20, 1840 for an "Improvement in the Mode of Communicating Information by Signals by the Application of Electro-Magnetism." It was the first of four patents that he would ultimately be granted for the telegraph.

The first telegraph message was sent on May 24, 1844. It was selected by the 18-year old daughter of the Commissioner of Patents, Anne Goodrich Ellsworth. The text was from the Bible (Numbers 23:23) "What hath God wrought". The last telegraph message was sent on July 12, 1999. It was the same message.

[3]See Stigler's law of eponymy which states that no scientific discovery is named after its original discoverer. Stigler's law was discovered by the sociologist Robert Merton.

1.2 George M. Dexter & Alexander Parris

Perhaps more than members of any other profession, architects live day in and day out at the intersection of science and art. They have to know the mathematics of stress and strain, the properties of steel and stone, the nature of storms and earthquakes, and current thinking in a dozen other scientific disciplines. At the same time their work must appeal to the aesthetic and artistic sensibilities of their clients as well as the rules and regulations—both artistic and structural—of the city in which it is to be executed.

One of the artistic works of George Minot Dexter (1802–1872) is all around you; he was the architectural engineer of the building in which you're standing. Dexter was born in Cambridge and graduated from Harvard in 1821. He spent the following ten years travelling in England, France and Germany during which time he purchased books for the Athenæum as well as our other architect-engineer, Alexander Parris.

Dexter is perhaps most well-known for the houses he designed in Pemberton Square. His architectural drawings—over 1,300 of them covering more than eighty-five different projects—are part of the Athenæum's collection.

One of the unique challenges facing the architects of designing the Athenæum was unusual contours of the plot on which it was to be built. The Granary Burying Ground which it abuts made something of a dent in the south-east side of the Athenæum's construction site.

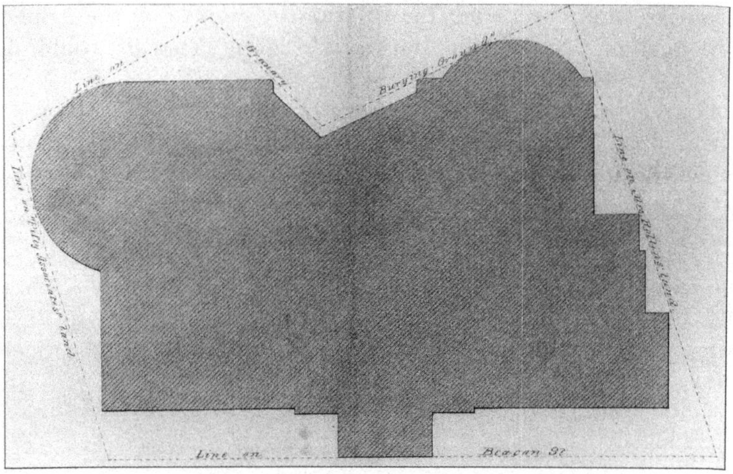

Figure 1.1: Plot and Footprint of the Boston Athenæum

As you can see the dent isn't in the building when it was finished. The solution to the problem required some three-dimensional artistic (and perhaps legal) thinking. The occupant of a grave may own the ground in which they are interred but, at least in this case, they don't own the air rights over the grave. The Athenæum was built over the graves in the indent.[4]

On December 1, 1840, Dexter was granted US Patent 1,875, for "Improvement in the Mode of Heating Buildings by Means of an Apparatus Consisting of Tubes for the Circulation of Hot Water Arranged in an Air-Chamber Adapted to the Same." The patent application starts out a as follows:

> To all whom it may concern: Be it known that I, George M. Dexter, of the city of Boston, in the State of Massachusetts, have invented a new and improved manner of constructing an apparatus for the warming of dwelling-houses and other buildings ... which may applied to various useful purposes, said improvement being specially applicable wherever a degree of artificial heat below that of boiling water is required or sufficient—such as, for example, as in the drying-rooms of laundries, bleacheries, manufactories of paper, gunpowder, and may other articles.

It is unlikely Dexter included his improvement in heating in his building at 10½ Beacon Street.

Alexander Parris (1780–1852) started out designing and building residential homes. He moved on to churches and public markets. The Cathedral Church of St. Paul and Quincy Market are both Parris designs. He ended up designing ropewalks, power magazines, and light houses.

Parris was closely associated with two utility patents although he isn't listed as a inventor on either of them. As the architect, consulting engineer, and construction supervisor of the ropewalk at the Charlestown Navy Yard he worked closely with both Daniel Treadwell and Ebenezer Lester. Lester's pendulum steam engine, X 4749½, and Treadwell's iron tail, X7997, were used in the ropewalk. Parris wrote a glowing report on Lester's engine for the *Franklin Journal.* The working relationship between Parris and Treadwell during the construction of the ropewalk is described by historian of technology Leslie Larson as follows:

> From the earliest conceptual stage for the Ropewalk to its completion Architect/Engineer Parris and Inventor Treadwell worked together closely on the planning and execution of what must be considered a collaborative effort.[9, p. 8]

---

[4]... graves of Sarah Gibson Cunningham (1666–1731) and Andrew Cunningham (1653–1735), in particular.

In addition to being an artist in brick and steel, Parris was also something of an artist in pen and paper. A small collection of his architectural drawings held by the American Antiquarian Society in Worcester and there forty-five of his drawings in his manuscript *Architectural Designs (1804–1810)*, held by the Athenæum.[12] A much larger collection can be found in his book *Plans of Buildings and Machinery Erected in the Navy-Yard Boston from 1830 to 1840*.

1.3 Chester Harding & Rembrant Peale

It's one thing to invent something and get a patent on your invention. It's something else altogether to realize any profit from it. James Watt invented and got a patent for his steam engine but it was his business partner, Matthew Boulton, who made the steam engine a commercial success. One might go so far as to say that Boulton invented the steam engine business.

There are two artists whose works are on display at the Athenæum who sought to play Boulton to inventions of others: Chester Harding (1792–1866) and Rembrant Peale (1778–1860).

Chester Harding painted the portrait of John Parker, Jr., and Rev. Thaddeus Mason Harris as well as the full-length portraits of Daniel Webster and John Marshall in the vestibule. Rembrant Peale painted the portrait of Samuel Eliot.

In his autobiography, Chester Harding writes:

> Early in the fall of this year I embarked in a new business. A mechanic had invented and patented a spinning-head, which was thought to be a great improvement upon the old plan. I accepted an offer he made me to sell the patent in the State of Connecticut. The only thing in the way of my making a fortune was the want of capital. However, "Where there's a will, there's a way." I soon contrived to get a horse and wagon, and five or six dollars in money, besides a quantity of essences, such as peppermint, tansy, wintergreen, etc. With this fit-out I launched forth into the wide world in pursuit of fortune.[6, pp. 8–9]

In the early 1800s, the artist, Rembrant Peale, owned and operated a museum in Baltimore[5] He illuminated his museum using lights burning carbureted hydrogen gas invented by Dr. Benjamin Kugler. Kugler had been granted patent X2572 on April 3, 1816, for a method of making the gas. In a lengthy (and self-promoting) article about Kugler's patent and how it stacks up against other ways of making fuel for lighting, Peale writes:

[5] . . . which can still be visited today.

Dr. Kugler's invention is not only that pitch, &c. can be advantageously used in the production of Gas; but that it furnishes a purer light, produces no offensive smell, requires no separation or purification through Lime water, or any other water, can be procured at less expense, and more easily, than by any other method.
[13, p. 529]

In fact, the gas lights themselves were something to marvel during a visit to Peale's museum.

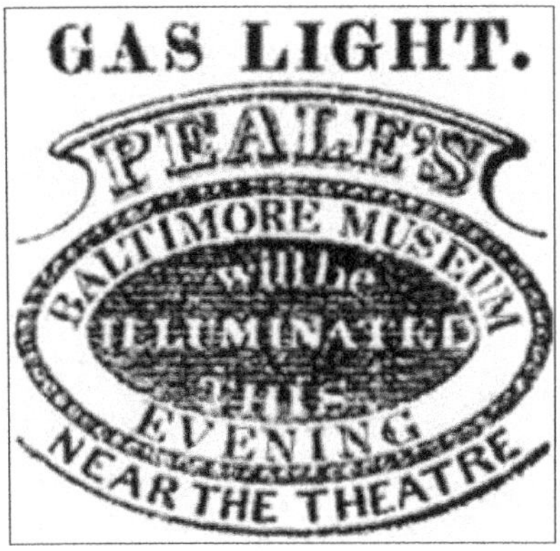

Figure 1.2: Gas Lights in Advertisement for Rembrant Peale's Museum

Peale was so taken with Kugler's gas lights that he licensed Kugler's patent and on February 5, 1817, incorporated Gas Light Company of Baltimore to light the streets of the city and supply the gas to its residents. But that's a story for another time.

It might be noted in passing that Rembrant Peale's father, Charles Willson Peale (1741–1827), was also an artist with inventive and entrepreneurial impulses. In 1804 he licensed the rights to patent X453 granted to John Isaac Hawkins on May 17, 1803, for a polygraph. Not to be confused with today's lie-detecting polygraph, the Hawkins-Peale polygraph was a way of producing additional copy of a handwritten document while it was being written; functionally what carbon paper is to a typewritten document. Being a great supporter of the latest technology, Thomas Jefferson also licensed the rights to practice Hawkins' invention and in the event worked with Peale to improve the device.

Chapter 2

More Stories of Art and Invention

2.1 Thomas Ball & Henry Dexter: Design Patents

When we think of patents we usually think about inventions that do things; Cyrus McCormick's reaper, Elias Howe's sewing machine, Eli Whitney's cotton gin, and so forth. These patents are called, for obvious reasons, utility patents. But there is another sort of patent that is an explicit connection art and invention. It is called a design patent and it protects an artistic expression.

Design patents are often applied for and granted for silverware designs, architectural flourishes, and unique aspects of human interfaces. Apple has received many design patents an example of which is D606,040 issued on December 15, 2009, for the iPod nano.

Another example of this sort of patent when it comes to the Athenæum's artists is the patent that the Boston sculptor Thomas Ball (1819–1911) applied for and received for his cabinet bust of the popular singer Jenny Lind. Ball was also granted design patents for other of his sculptures.

Henry Dexter (1806–1876) was granted a design patent for his full-size statue of General Joseph Warren.

2.2 Hiram Powers: Sculpting Tools and 3-D Printing

It is said that necessity is the mother of invention. In pursuing their profession, artists have necessities just like everybody else. The sculptor Hiram Powers (1805–1873) invented tools for modeling in plaster as well as a machine for cloning three-dimensional objects.

The Metropolitan Museum of Art's biography of Powers notes:

> Powers continued to model portrait busts throughout his career, a genre in which he excelled, completing some 150. George Washington, his most popular, was translated to marble at least thirty-six times. In order to fulfill orders for East Coast commissions that had been executed in the preliminary materials of plaster and clay, he learned to carve marble in 1838, advised by the assistant of fellow Florentine expatriate sculptor Horatio Greenough. As Powers' practice expanded and his popularity grew, he relied on a workshop of trained craftsmen to assist him in molding, casting, and carving his works. He invented time-saving tools and machines that streamlined the sculptural process for himself and his assistants. To complement his sculptures, he designed accompanying pedestals that were fabricated in his studio.

One of Powers' most famous sculptures, the life-sized Greek Slave, was reproduced using three different cloning methods during the Smithsonian's 2015 exhibition of Powers' work, Measured Perfection: Hiram Powers' Greek Slave. The exhibition featured not only Powers' sculptures but also the tools that he invented to create them. The write-up of the exhibition says:

> This one-gallery exhibition reveals the inner workings of the studio of Hiram Powers, who was among the most innovative sculptors of the nineteenth century, eagerly adapting long-standing sculpture traditions to new technologies of his age.

Powers applied for a design patent on the Greek Slave statue on July 7, 1849, but there is no record of the patent having been granted.

2.3 Wax Painting and Klein Blue

We usually think of technical innovation as being somehow socially more worthy of patent protection than artistic innovation. The quid pro quo of patent protection is that the inventor gets exclusive commercial use of the innovation for a fixed period of time in exchange for describing the innovation in sufficient detail that anybody can make use of it afterwards. The alternative is that the invention dies with the inventor and its use is lost to society forever ... or at least until it is re-invented. It's hard to apply this quid pro quo to a sculpture or a symphony or a dance.

The French philosopher, writer, and art critic Denis Diderot (1713–1784) disagreed with this line of reasoning. Diderot argued that artists should receive vigorous legal protection for their works while being able to use freely use whatever technology was available. Inventors were mere technicians.

Diderot's views were in play in the middle of a controversy over the invention of a supposedly new way of painting called encaustic or hot wax painting. The story is well-told in an article by Lilane Hilaire-Pérez, "Diderot's Views on Artists' and Inventors' Rights: Invention, Imitation and Reputation" [7].

Diderot's argument gets a bit convoluted when it is the artists who are inventing the technology. The controversy was between two French artists both of whom claimed to have invented hot wax painting. As it turned out the technique was used to decorate Egyptian mummies. The term "prior art" takes on a dual meaning in this case.

There are, in fact, other instances of artists seeking priority fame and legal protection for the invention of artistic media in distinction to artistic tools as in the case of Powers' files above. The French painter Yves Klein (1928–1962), for example, sought protection for color that is now known as International Klein Blue. Klein's US application was rejected but he did receive a French patent, FR1258418, for a method of painting using models covered in his blue. This method of painting is said to be the inspiration of Boston's Blue Man Group.

2.4 The Kazoo, the Moog, and the Theremin

One need only gaze upon a French horn or a contrabassoon to appreciate that a prodigious amount of technical knowledge and creativity goes into the invention of musical instrumentation.[1] One might posit that musical instruments are only invented by musicians. Who, after all, would care more about a new way of making music than somebody who was already making an old way? And, indeed, many musical instruments have traceable genealogies that wind through virtuosos, composers, and orchestras. Many, but not all.

[1] ... except perhaps for the conductor's baton which was invented, if that be the term, by Louis Spohr in the 1820's.

George Smith, Robert Moog, and Léon Theremin, the inventors of the kazoo, the Moog synthesizer, and the theremin respectively, certainly were interested in music. Who isn't, after all? But they came to be thought of as musicians only after they had invented their musical instrument. All three were inventors who became artists.

Very few inventions are of something wholly new under the sun. The steam engine and the transistor are examples of inventions that took giant steps but almost all inventions take baby steps. The steam locomotive and the integrated circuit chip are, it is obvious and widely-acknowledged, the products of thousands of inventive insights.

Musical instruments are no exception to this rule. Just as an example, Boston was the home of the leading manufacturers of piano-fortes in the nineteenth century. It is not surprising, then, that Boston residents were granted numerous patents for improvements in Bartolomeo Cristofori's eighteenth century invention.

Appendix A

Tables

Number	Date	Inventor	Title
	1416	Francisco Petri	Fulling Wool. First Patent.[1]
	1421	Filippo Brunelleschi	Boat called "Il Badalone." (often cited as 1st patent)
	1449	John of Utynam	Stained Glass. First English Patent.
	1641	Samuel Winslow	Making Sails. First New World Patent.[2]
X1	July 31, 1790	Samuel Hopkins	Making Pot & Pearl Ashes. First US Utility Patent.
US 1	July 13, 1836	John Ruggles	Locomotive Steam-Engine. First Patent after renumbering.
D1	Nov. 9, 1842	George Bruce	Type font. First US Design Patent.

Table A.1: Some Firsts in the History of Patents

[1] There is not a consensus in the legal community regarding the first ever patent.
[2] . . . issued by the Massachusetts General Court

Number	Date	Inventor	Title
X2843	Oct. 3, 1817	Samuel Morse[3]	Flexible-piston man-powered water pump for fire engines
US 1,647	Jun. 20, 1840	Samuel Morse	Improvement in the mode of communicating information by signals by the application of electro-magnetism
US 3,316	Oct. 25, 1843	Samuel Morse	Method of introducing wire into metallic pipes
US 4,453	Apr. 11, 1846	Samuel Morse	Improvement in electro-mechanical telegraphs
US 6,420	May 1, 1849	Samuel Morse	Improvement in electric telegraphs
US 1,875	Dec. 1, 1840	George M. Dexter	Improvement in the Mode of Heating Buildings by Means of an Apparatus Consisting of Tubes for the Circulation of Hot Water Arranged in an Air-Chamber Adapted to the Same
US 10,088	Oct. 4, 1853	Hiram Powers	File and Rasp
British 1592 of 1857	June 6, 1857	Hiram Powers	An Improved Machine for Punching, Stamping, or Cutting Metals and other Substances
British 1221 of 1860	May 17, 1860	Hiram Powers	Improvements in vices, part of which improvements is [sic] applicable to screw wrenches, spanners, turning lathes, and other tools and machines
US 31, 476	Feb. 19, 1861	Hiram Powers	Punching Machine

Table A.2: Utility Patents of Boston Artists

[3] . . . with brother Sidney and Samuel Findley

Number	Date	Inventor	Title
D369	Apr. 29, 1857	Thomas Ball	Design for a Bust of Jenny Lind
D548	Apr. 19, 1853	Thomas Ball	Design for a Bust of Daniel Webster
D590	Aug. 9, 1853	Thomas Ball	Design for a Statuette of Daniel Webster
D894	Jun. 9, 1857	Thomas Ball	Design for a Bust on Napoleon Bonaparte
D919	Aug. 4, 1857	Henry Dexter	Design for a Statue of General Warren

Table A.3: Design Patents Granted to Boston Artists

Number	Date	Inventor	Title
X3131	Sept. 6, 1819	Thomas Blanchard	Turning Irregular Forms
X4749½	May 14, 1827	Ebenezer Avery Lester	Pendulum Steam Engine
X7997	Feb. 5, 1834	Daniel Treadwell	Iron Tail
US 6,469	May 22,1849	Abraham Lincoln	Bouying of Vessels over Shoals
US 161,739	Apr. 6, 1875	Alexander Graham Bell	Transmitter and Receiver for Electric Telegraphs
US 174,465	Mar. 7, 1876	Alexander Graham Bell	Telegraphy[4]
US 121,992	Dec. 19, 1871	Samuel Clemens	Improvement in Adjustable and Detachable Straps for Garments
US 140,245	Jun. 24, 1873	Samuel Clemens	Improvement in Scrap-Books
US 324,535	Aug. 18, 1885	Samuel Clemens	Memory Builder Game
US 606,040D	Dec. 15, 2009	Bartley K. Andre et al	Media Device[5]

Table A.4: Patents Relevant to the Up Close Tour

[4] ...the first of eleven. This patent is said to be the most valuable patent ever issued.
[5] In particular, the iPod nano.

Inventor	Invention
William Morton	surgical anesthesia
James Bogardus	cast-iron construction
Samuel Colt	revolving pistol
Cyrus Hall McCormick	mechanical reaper
Joseph Saxton	coal-burning stove, hydrometer, ever-pointed pencil
Charles Goodyear	vulcanization of rubber
Peter Cooper	railway locomotive
Jordan Lawrence Mott	coal-burning cooking stove
Joseph Henry	electromagnet design
Eliphalet Nott	efficient heat conduction for stoves and steam engines
John Ericsson	armored turret warship
Frederick Sickels	steam-engine gear and steering device for ships
Samuel F. B. Morse	electric telegraph
Henry Burden	horseshoe manufacturing machine
Richard March	rotary press
Erastus Bigelow	power loom for carpets
Isaiah Jennings	threshing machine, repeating gun, friction match
Thomas Blanchard	irregular turning lathe
Elias Howe	sewing machine

Table A.5: Inventors in Schusselle's *Men of Progress* (left to right)

Appendix B

Illustrations

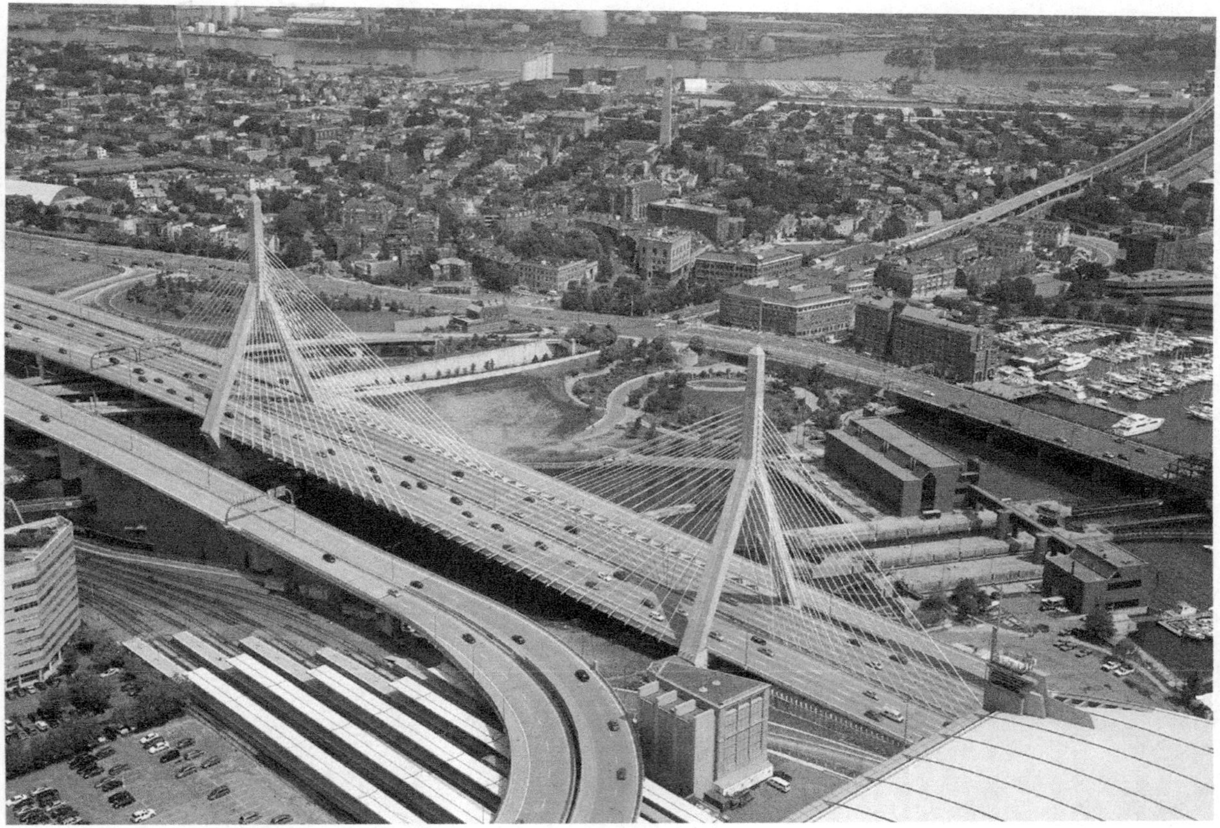

Figure B.1: Leonard P. Zakim Bunker Hill Memorial Bridge

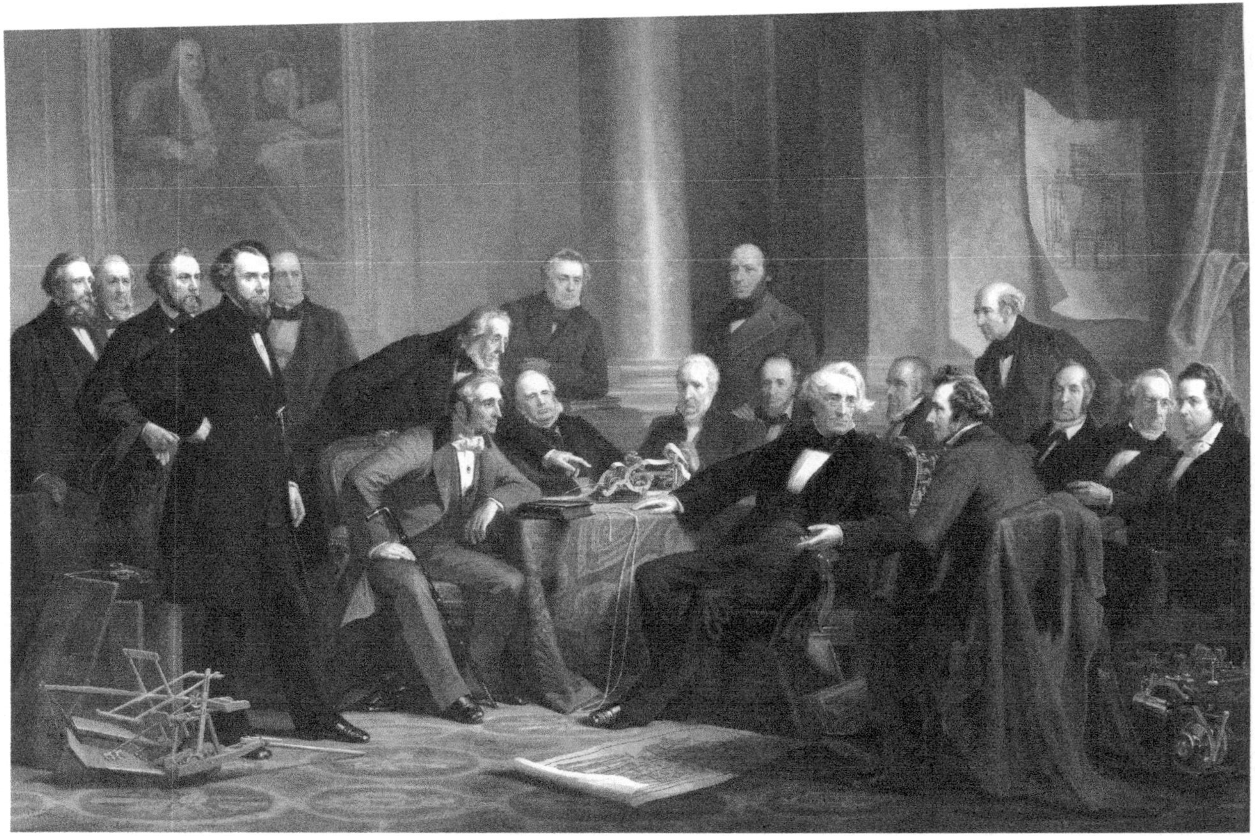

Figure B.2: Men of Progress (1857) by Christian Schussele

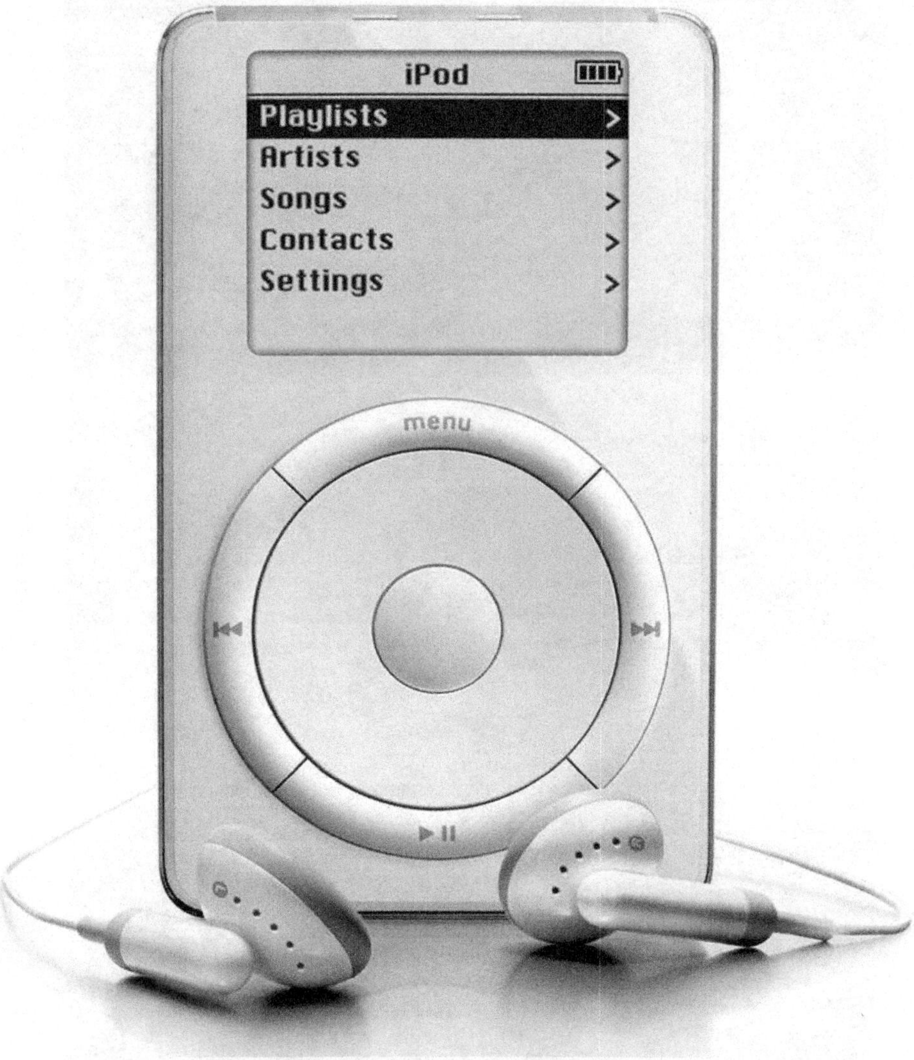

Figure B.3: First Generation iPod

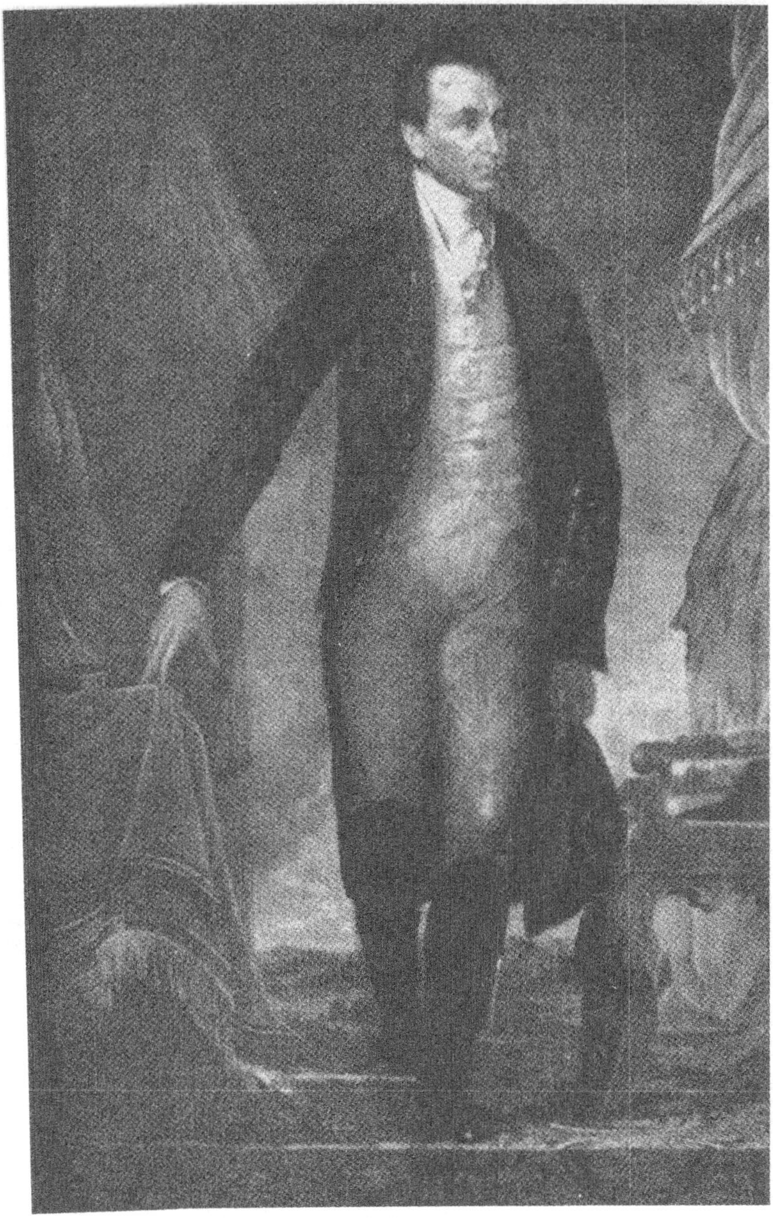

Figure B.4: Morse's Full-Length Portrait of James Monroe

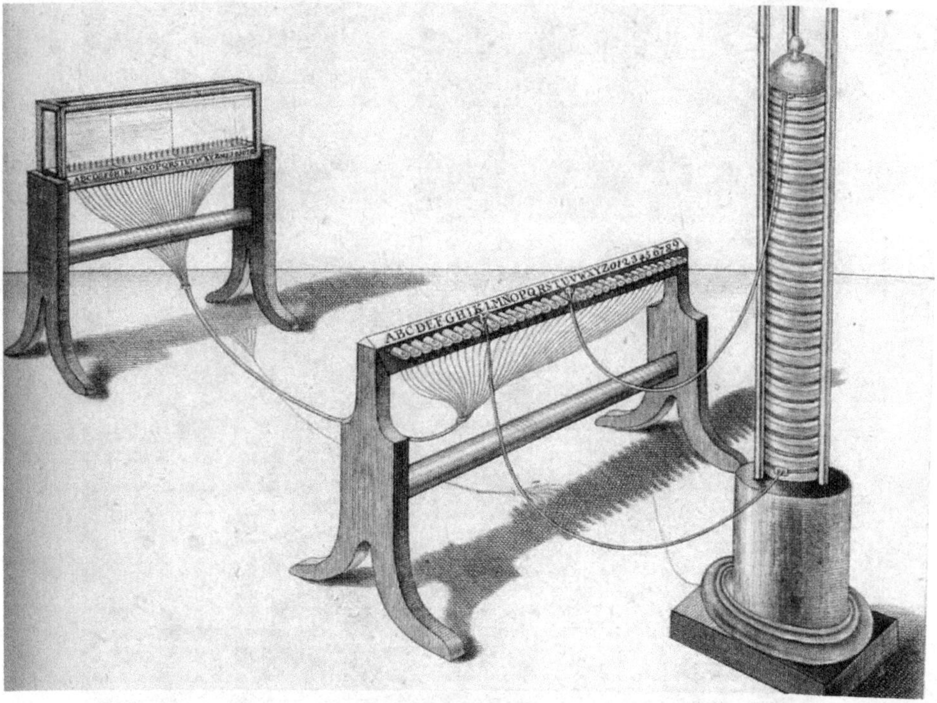

Figure B.5: Transmitting Letters Electrically Using Bubbles

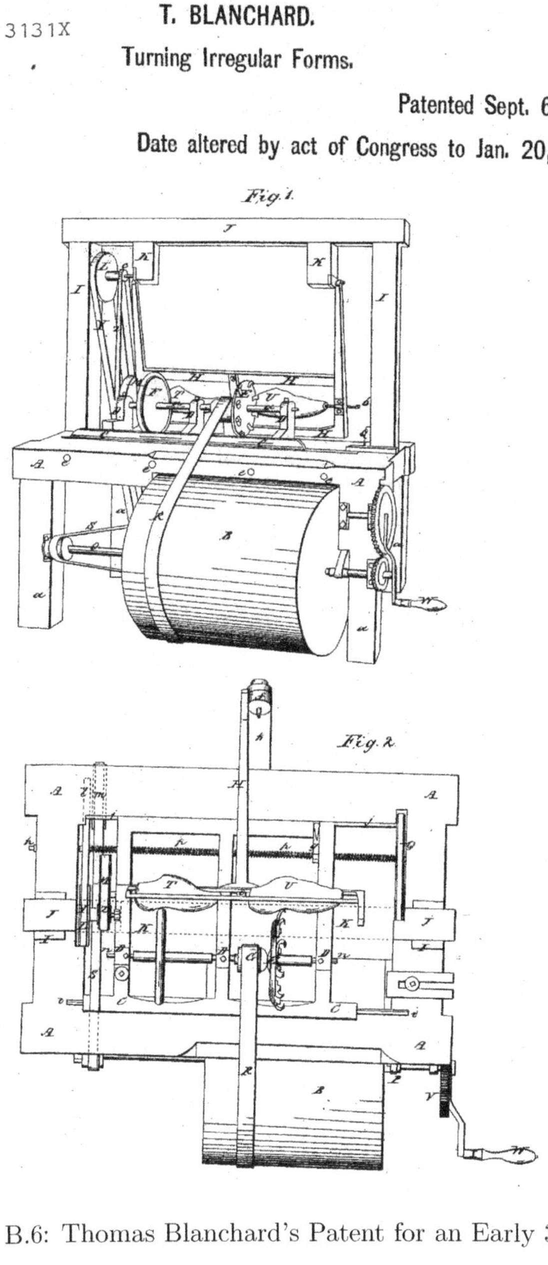

Figure B.6: Thomas Blanchard's Patent for an Early 3D-Printer

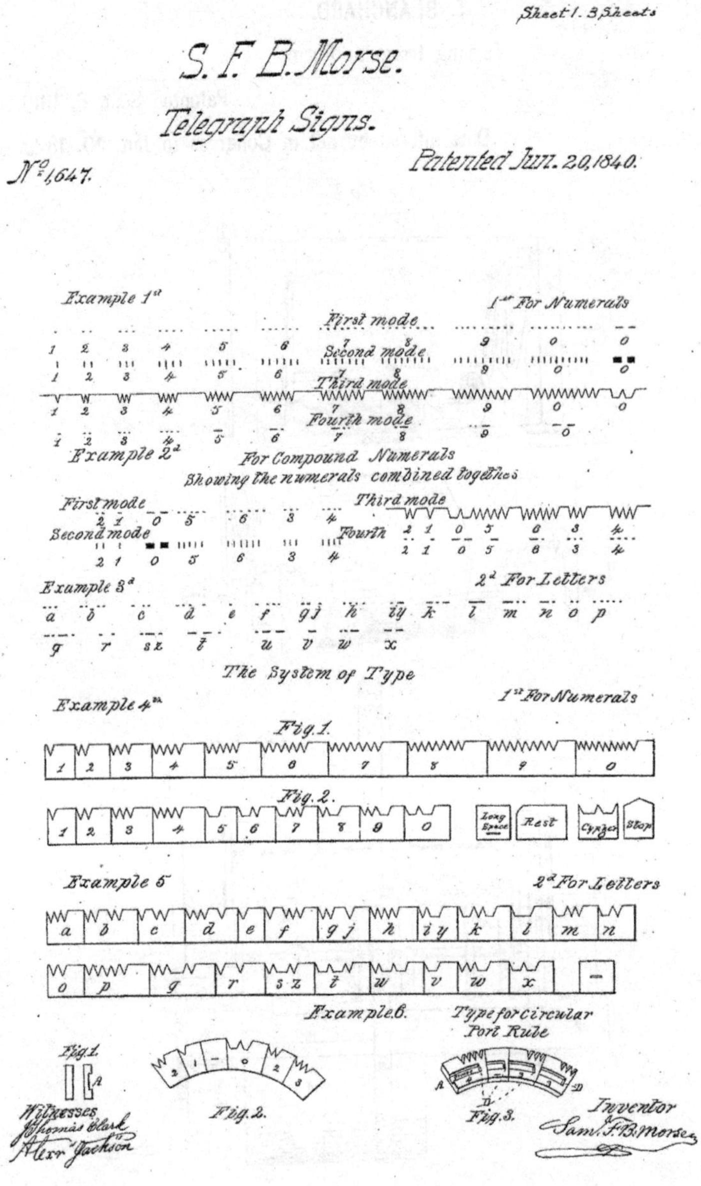

Figure B.7: Morse's First Patent for the Telegraph

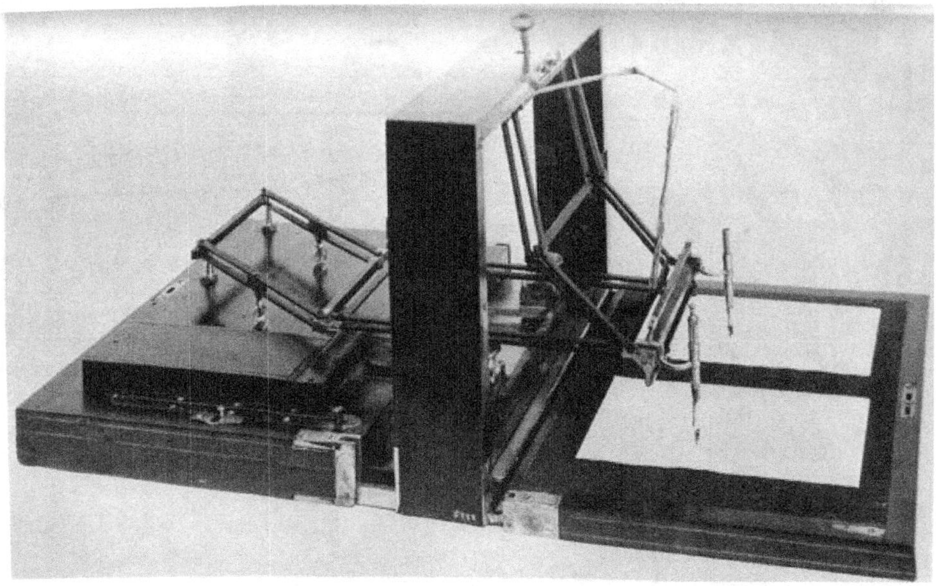

Figure B.8: The Hawkins & Peale Polygraph

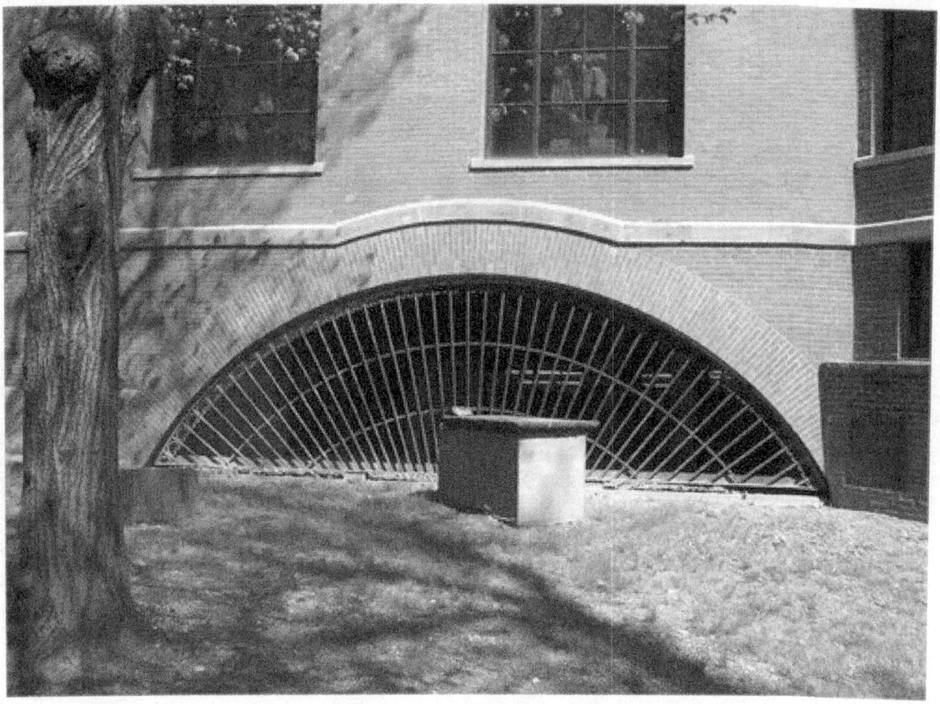

Figure B.9: Arch over the Cunninghams' Graves

G. M. DEXTER.

Steam Heater.

No. 1,875. Patented Dec. 1, 1840.

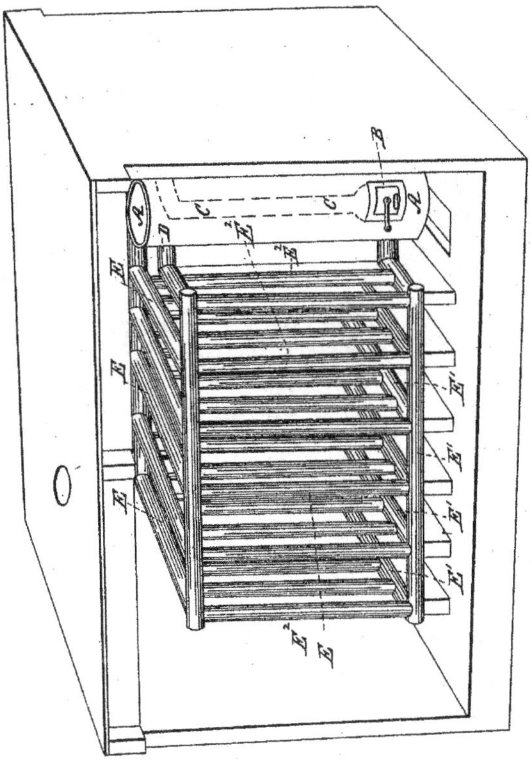

Figure B.10: George M. Dexter's Heater Patent

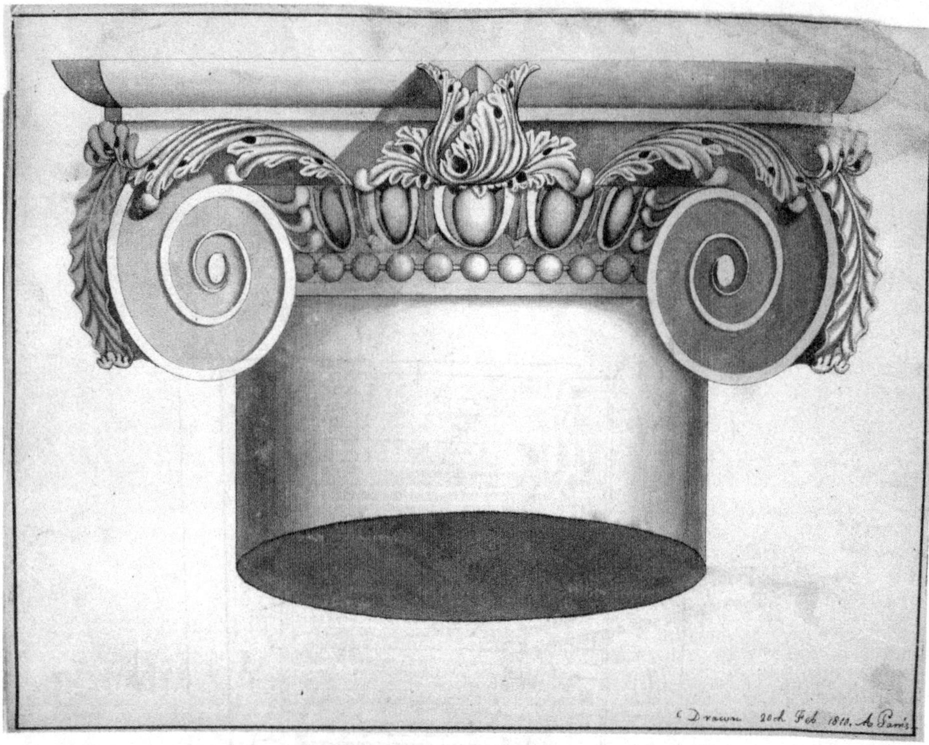

Figure B.11: Early Drawing by Alexander Parris

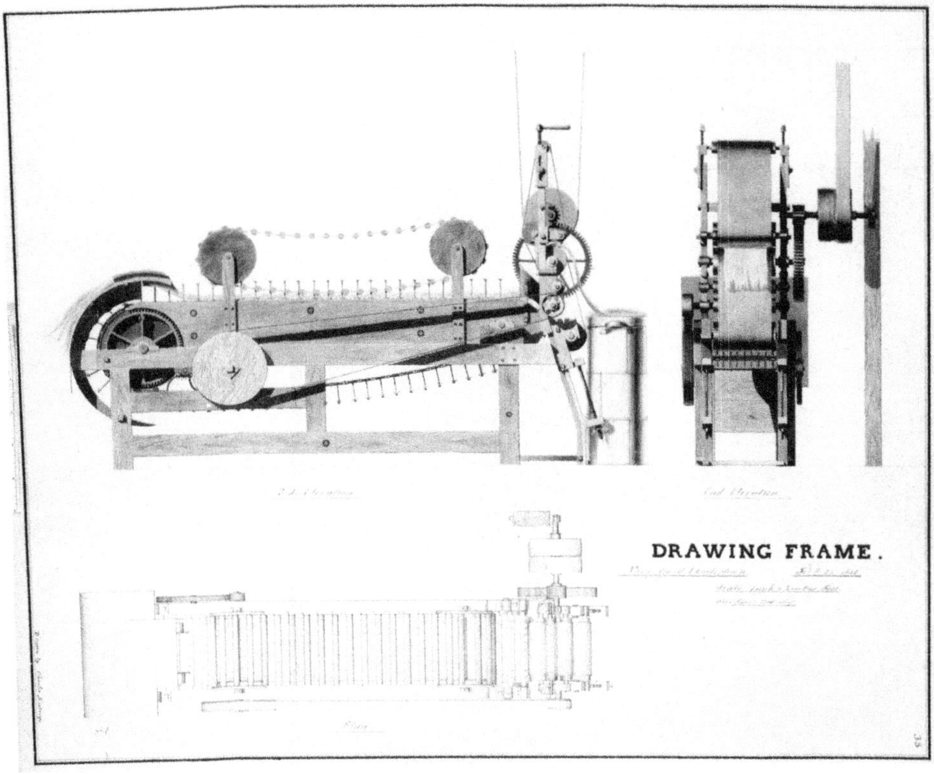

Figure B.12: Later Drawing by Alexander Parris

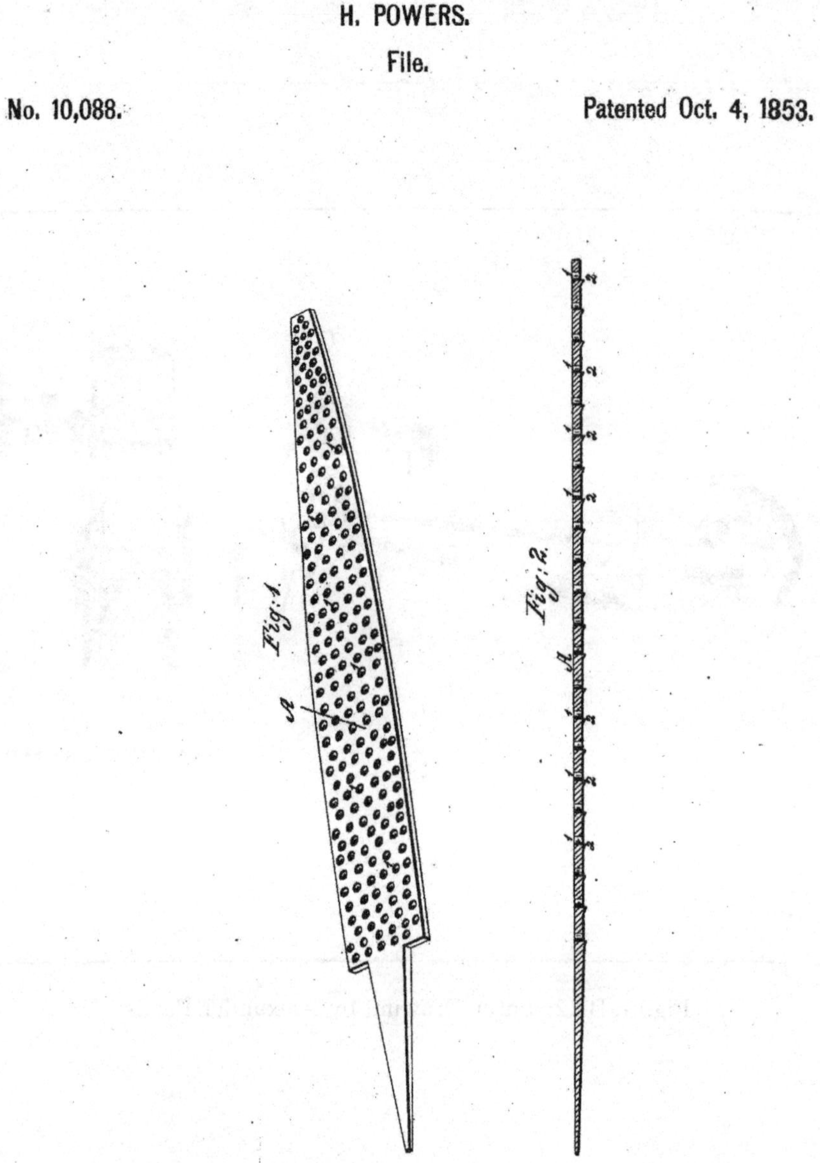

Figure B.13: Hiram Powers' Patent for a File

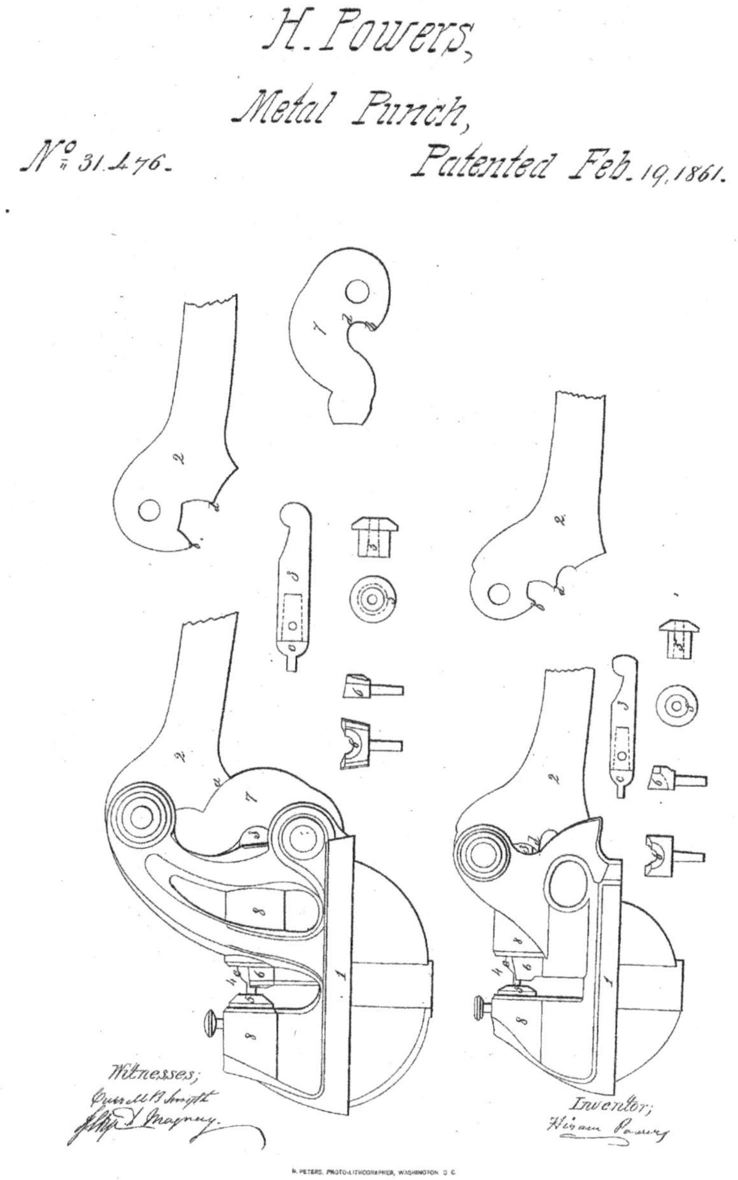

Figure B.14: Hiram Powers' Patent for Sculpting Tools

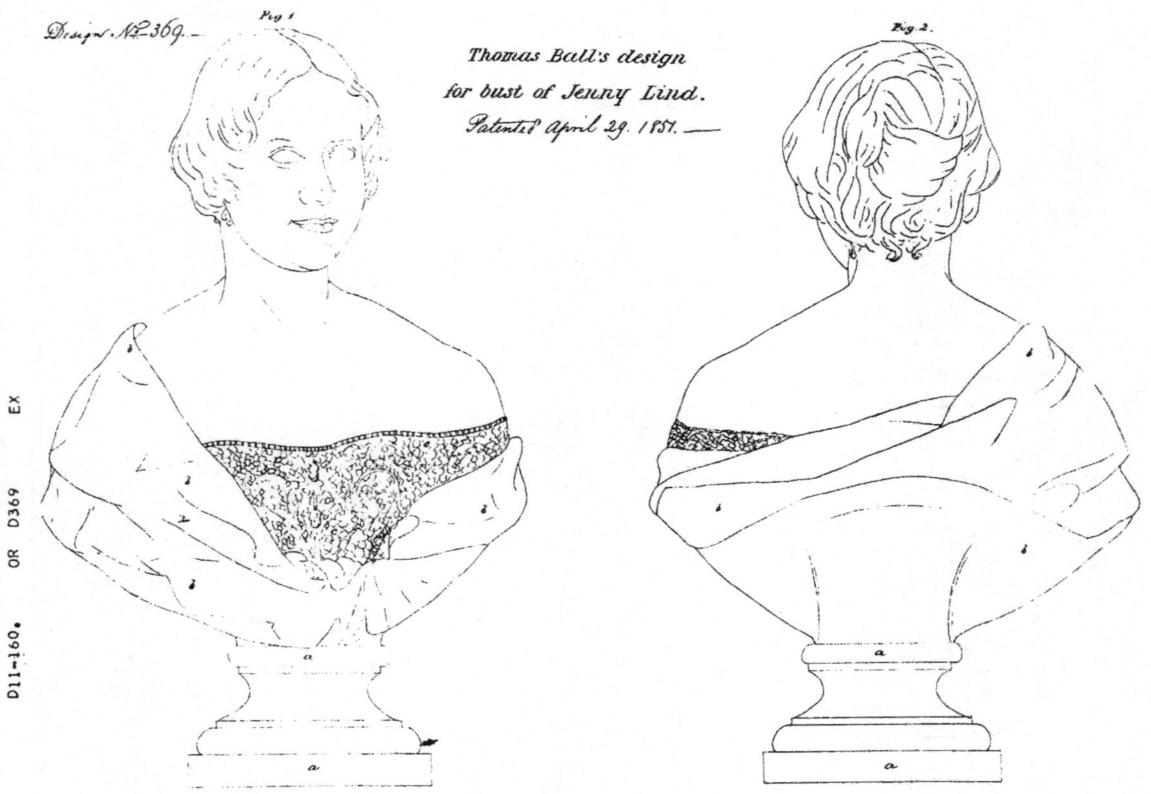

Figure B.15: Ball's Design Patent for Cabinet Bust of Jenny Lind

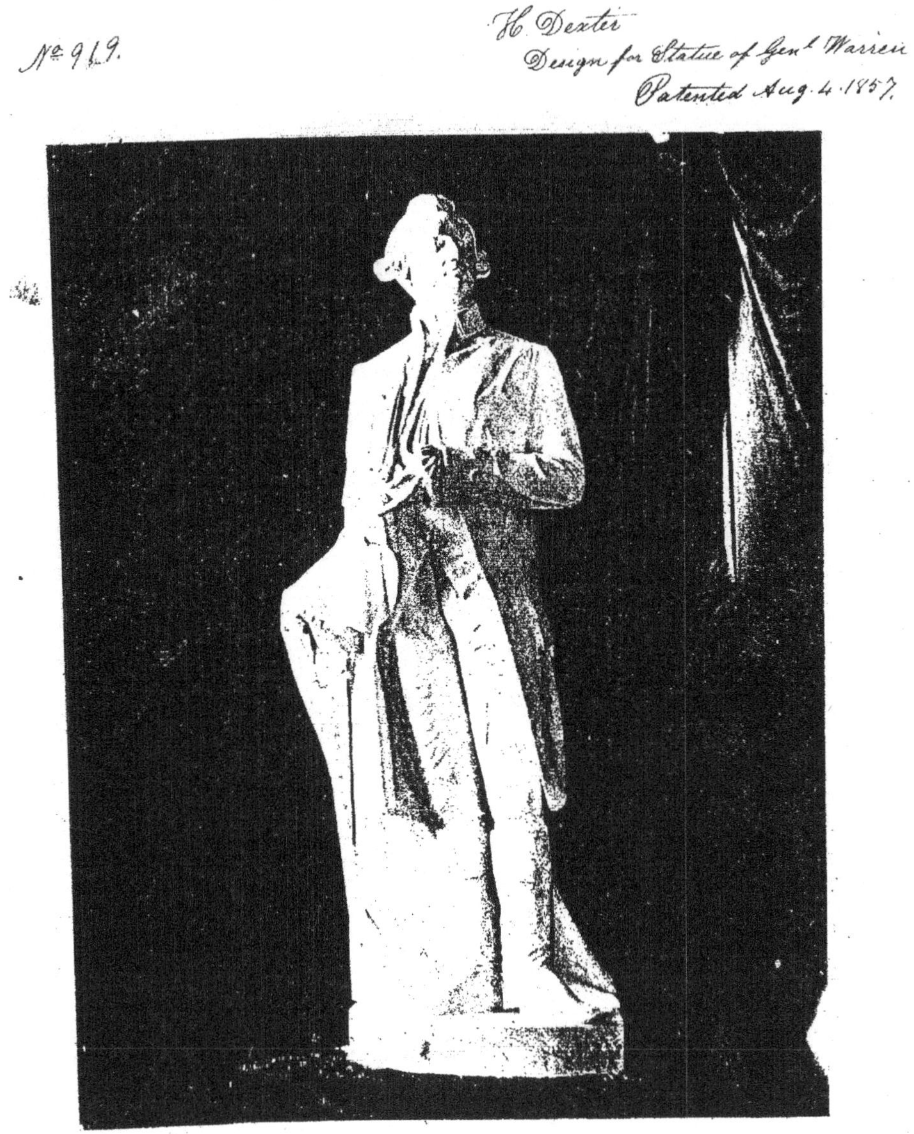

Figure B.16: Henry Dexter's Design Patent for Statue (drawing)

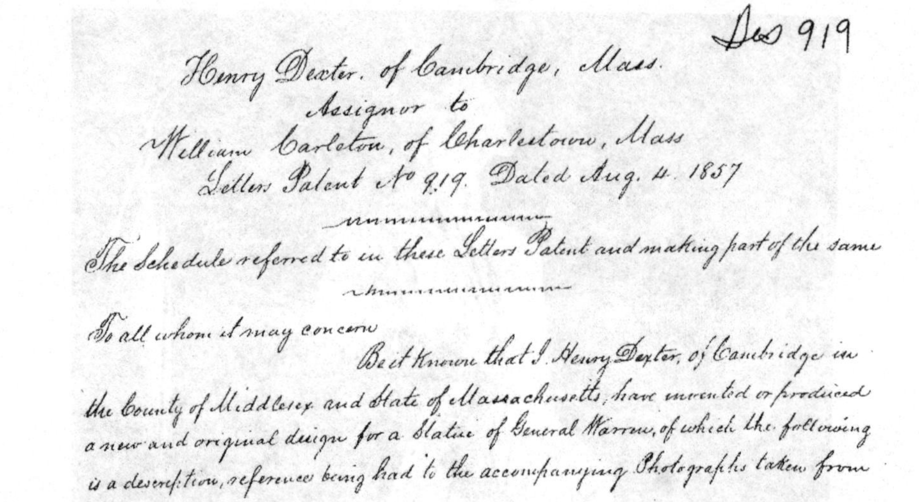

Des 919

Henry Dexter. of Cambridge, Mass.
Assignor to
William Carleton, of Charlestown, Mass
Letters Patent No 919. Dated Aug. 4. 1857

The Schedule referred to in these Letters Patent and making part of the same

To all whom it may concern

Be it Known that I, Henry Dexter, of Cambridge in the County of Middlesex and State of Massachusetts, have invented or produced a new and original design for a Statue of General Warren, of which the following is a description, reference being had to the accompanying Photographs taken from the Statue.

The said Photographs represent a Statue of General Warren of full length, with military boots and military frock coat, the vest buttoned up at the throat; with the left foot slightly advanced, the left hand in a position crossing the body, the right hand resting on the hilt of a sword the point of which rests on a pedestal. The figure is represented without a hat with the hair curled at the sides of the head

A.T.M Witnesses
Sam. Cooper
Thos R. Roach

260

Henry Dexter

Figure B.17: Henry Dexter's Design Patent for Statue (text)

(12) **United States Design Patent** (10) **Patent No.: US D476,424 S**
Shanahan (45) **Date of Patent: ** Jun. 24, 2003**

(54) **BUILDING STRUCTURE FACADE**

(76) Inventor: **Paul J. Shanahan**, 105 E. Helen, Palatine, IL (US) 60067

(**) Term: **14 Years**

(21) Appl. No.: **29/167,142**

(22) Filed: **Sep. 10, 2002**

(51) **LOC (7) Cl.** **25-02**
(52) **U.S. Cl.** **D25/59**
(58) **Field of Search** D25/59; 52/415, 52/311.1, 315

(56) **References Cited**

U.S. PATENT DOCUMENTS

D217,857 S * 6/1970 Miller D25/59
D425,208 S * 5/2000 Allison D25/59

* cited by examiner

Primary Examiner—Doris Clark
(74) *Attorney, Agent, or Firm*—Wallenstein & Wagner, Ltd.

(57) **CLAIM**

The ornamental design for a "building structure facade", as shown and described.

DESCRIPTION

FIG. **1** is a perspective view of the building structure facade in relation to the remainder of the building;
FIG. **2** is a perspective view of the building structure facade;
FIG. **3** is a rear view thereof;
FIG. **4** is a top view thereof;
FIG. **5** is a right side view thereof; and,
FIG. **6** is the left side view thereof.
The bottom of the building structure facade, which is generally not viewed or observed, is plain and unornamented.
Features shown in broken lines in the drawings are not part of the claim.

1 Claim, 5 Drawing Sheets

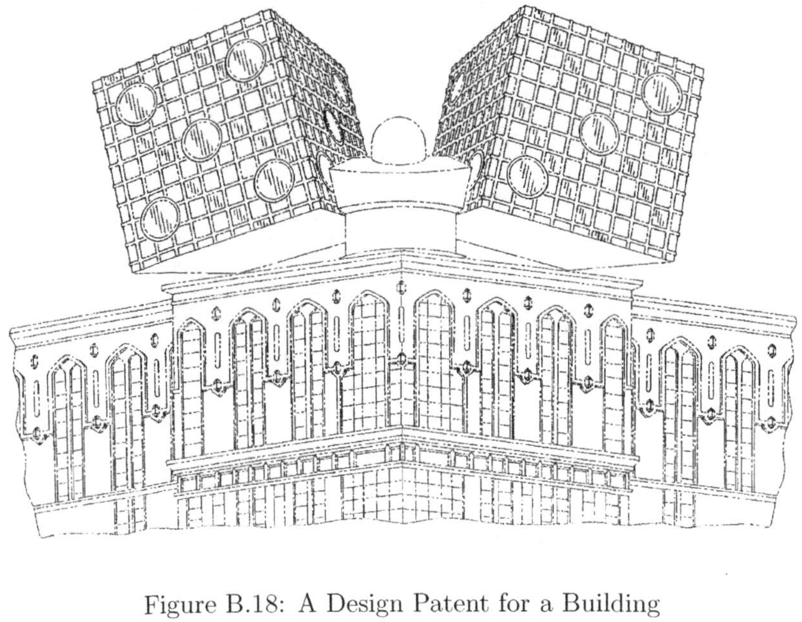

Figure B.18: A Design Patent for a Building

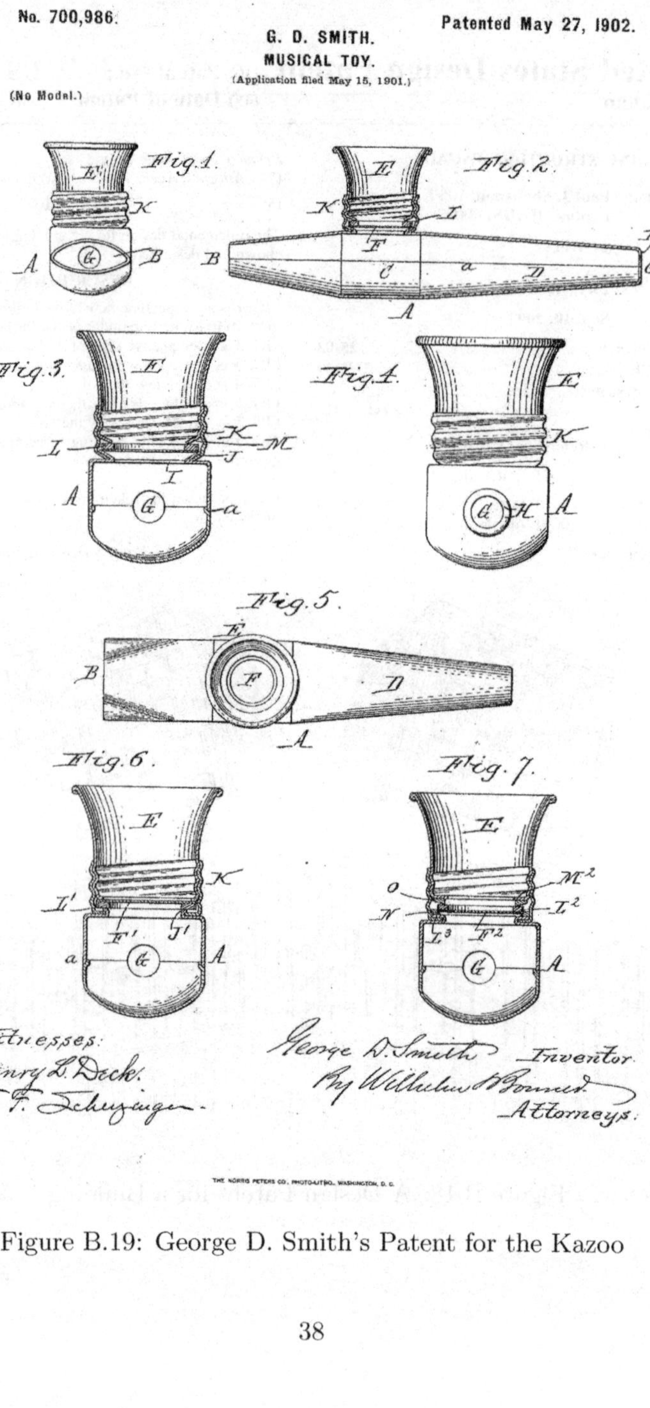

Figure B.19: George D. Smith's Patent for the Kazoo

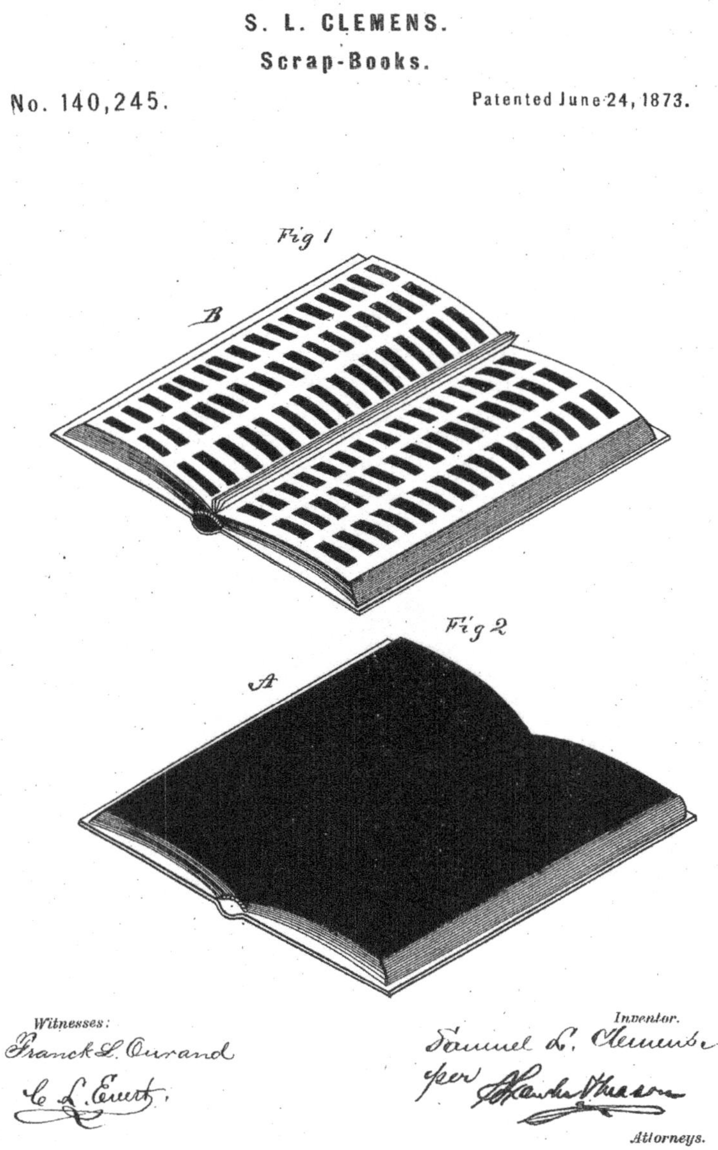

Figure B.20: Mark Twain's Patent for a Scrapbook

Further Reading

Entries for books in the Boston Athenæum's current collection include callnumber and Athena bibId within square brackets and separated by an at sign (@).

[1] Anonymous. *Copies of the claims and patents of S.F.B. Morse : as shown in correspondence with the Patent Commissioner, &c. Atlantic, Lake and Mississippi Telegraph, &c.* s.n, S.l, 1 ed., 1848. [NV .M83 c @ 231512].

[2] Thea Burns. *The invention of pastel painting.* Archetype Publications, London, 1 ed., 2007. [NC880 .B87 2007 @ 451183].

[3] Dieter Daniels, editor. *Artists as Inventors–Inventors as Artists.* Hatje Cantz, Berlin, 2008.

[4] Charles Gobrecht Darrach Darrach. Christian Gobrecht, Artist and Inventor. *The Pennsylvania Magazine of History and Biography*, 30(3):355–358, 1906.

[5] Sarah Kate Gillespie. *Samuel F. B. Morse and the daguerreotype : art and science in American culture, 1835–55.* n.s., n.s., 1 ed., 2006. [TR365 .G547 2006 @ 460544].

[6] Chester Harding. *My egotistigraphy. Prepared for his family and friends, by one of his children.* John Wilson Press, Cambridge Mass., 1 ed., 1866. [U9 .H213 .m @ 249973].

[7] Liliane Hilaire-Pérez. Diderot's Views on Artists' and Inventors' Rights: Invention, Imitation and Reputation. *The British Journal for the History of Science*, 35(2):129–150, 2002.

[8] Oliver W Larkin. *Samuel F. B. Morse and American democratic art.* Little, Brown, Boston, [1st ed.] ed., 1954. [65 .M8391 .l @ 136940].

[9] Leslie Larson. *The Ropewalk at the Charlestown Navy Yard: A History and Reuse Plan.* Boston Redevelopment Authority, Boston, 1987.

[10] Jean Lipman. *Rufus Porter rediscovered : artist, inventor, journalist, 1792-1884 / Jean Lipman ; foreword by Gerard Piel, foreword to the 1980 edition by John I.H. Baur.* C.N. Potter : Distributed by Crown, New York, 1 ed., 1980. [CT275. P5966 L56 1980 @ url].

[11] Samuel Finley Breese Morse, Peter John Brownlee, and Jean-Philippe Antoine. *Samuel F.B. Morse's Gallery of the Louvre and the art of invention / edited by Peter John Brownlee ; with essays by Jean-Philippe Antoine, and twelve others.* n.s., n.s., 1 ed., 1982. [ND237.M75 A65 2014 @ 512968].

[12] Alexander Parris. *Architectural designs, 1804–1810.* n.s., n.s., 1 ed., 2012. [.L333 @ 40183].

[13] Rembrant Peale. On Gas Light. *The Portico, A Repository of Science and Literature*, I:523–530, January-June 1816.

[14] Kenneth Silverman. *Lightning man : the accursed life of Samuel F.B. Morse / Kenneth Silverman.* Alfred A. Knopf, New York, 1st ed ed., 2003. [CT275.M67536 S54 2003 @ 402270].

[15] Paul J. Staiti. *Samuel F. B. Morse.* Cambridge University Press, Boston, 1990.

[16] Paul J. Staiti and Gary A. Reynolds. *Samuel F. B. Morse / essays by Paul J. Staiti and Gary A. Reynolds.* Grey Art Gallery and Study Center, New York University, New York, 1 ed., 1982. [ND237.M75 A4 1982 @ 20272].

[17] Robert Thill. Intellectual Property: A Chronological Compendium of Intersections between Contemporary Art and Utility Patents. *Leonardo*, 37(2):117–124, 2004.

Index

Made in the USA
Monee, IL
04 November 2026

67375598R00030